"The contemplation of the u
only to the adoration of our
better responsibility for our ac
of our damage to the envii _____ ___ misuse of
natural resources. I hope this selection of prayers helps
shape the praying heart of the Church and enliven our
discipleship."

— **Robert Atwell**, *Bishop of Exeter and*
Chair of the Liturgical Commission

—

"This wonderful book fills a much-needed gap in the
liturgical imagination of the contemporary Church.
Prayer for the earth has never been more needed, and
yet churches, prayer groups and intercession leaders
lack the resources to meet this need. Rob Kelsey has
produced a comprehensive, clear and well-planned
prayer resource, rooted both in theology and in a
knowledge of the practical exigencies of environmental
care and global stewardship. His introduction elegantly
makes the entirely compelling case that prayer for the
earth is not just desirable but necessary as an authentic
expression of Christian faith. This is a resource that
ought to be widely used and deeply appreciated in the
churches."

— **Carmody Grey**, *Assistant Professor of*
Catholic Theology at Durham University

"Prayerful thanksgiving for creation, as well as lamenting how humans are harming God's gift to us, are imperatives for leading us to hope-filled action. Rob Kelsey's beautifully crafted prayers, together with his clear outline for including environmental joys and concerns in our private devotions and public worship, are a must have resource for the Church at a time of climate and biodiversity emergency."

— **Graham Usher**, *Bishop of Norwich and ecologist*

Praying for the Earth

*Remembering the Environment in
our Prayers of Intercession*

— ROB KELSEY —

Sacristy
Press

Sacristy Press
PO Box 612, Durham, DH1 9HT

www.sacristy.co.uk

First published in 2021 by Sacristy Press, Durham.

Sacristy Limited, registered in England & Wales, number 7565667

British Library Cataloguing-in-Publication Data
A catalogue record for the book is available from the British Library

ISBN 978-1-78959-135-4

Contents

Foreword

Praying for the Earth is central to the Christian vocation to care for God's creation. It is of a different order from the other crises that currently face us in this age of pandemic. Change seems to be coming at us faster than before COVID-19. It is already clear that the economic crisis is going to be deep and recovery will not be quick. Black Lives Matter has highlighted the need for social change and greater inclusion, addressing the issues of history by the way we live in the present. There are big opportunities here if we face these challenges well. What are the beliefs and values that will shape us individually and as a society? What sort of people do we want to be?

The climate and environmental crisis is on a different scale, an altogether bigger and more existential threat, and opportunity. Scientists, economists and politicians have come to a broad agreement about the serious implications of climate change and environmental destruction caused by the way we human beings live. We are developing technical solutions and making progress towards decarbonization that would have been inconceivable even ten years ago.

There are opportunities here, too, and they need to be integrated into the responses to the other crises we currently face. Building Back Better has the potential to reorientate the economy and create Green jobs at this time of enormous social change. Like the pandemic, the climate and environmental crises are teaching us that we must live sustainably, locally and globally.

If we are to achieve the aims of the 2015 Paris Agreement and limit the global rise in temperature by 2 degrees Centigrade above pre-industrial levels by 2050, rapid change is needed in this decade to 2030. Yet there are real doubts about whether we are moving fast enough. What is required is not just a technical fix but spiritual change.

Our era is characterized as the Anthropocene because the most important changes happening to the planet are caused by the way we human beings live. We can't go on living as though the gift of creation is all to do with us. The challenge to us is spiritual. What is our relationship with God, one another and all creation?

The Anglican Communion has Five Marks of Mission:

- To proclaim the Good News of the Kingdom.
- To teach, baptize and nurture new believers.
- To respond to human need by loving service.
- To seek to transform unjust structures of society, to challenge violence of every kind and to pursue peace and reconciliation.
- To strive to safeguard the integrity of creation and sustain and renew the life of the earth.

To safeguard the integrity of creation is not an optional extra. It was added to the other four because they don't make sense in our world without it. It connects with people who don't go to church. It matters profoundly to the young. It is a missionary imperative.

In February 2020 the Church of England's General Synod made a decision for the Church to become Carbon Net Zero by 2030. Those who made the proposal argued that we should seek to move faster than the 2050 goal of the Paris UN Climate Change agreement. They wanted the Church to pick up the pace, demonstrate the urgency and for God's sake give a lead to the wider society. To achieve this will take practical action, but it won't happen unless we pray for the earth and make this commitment our central concern.

Prayer helps us to pay attention and really notice the beauty of creation, value it for itself, not just for the use we can make of it, and give thanks for it. In prayer we find ourselves responding to our failures, seeking forgiveness for having misused and abused God's good creation and, instead of being paralysed by the enormity of the task, turn things round, so that we live better in ways that value what God has given us, so as to sustain and renew life.

In what Jesus said about prayer, he taught us to be simple and direct, and to pray for what we want. By articulating what we want we are seeking to align our lives with God: "Thy will be done on earth as it is in heaven". Intercessory prayer clarifies our desires and connects what we want with what we do. The spiritual becomes the driver for change. It is in prayer that we find the will for fundamental change, individually and together.

Praying for the Earth is so important. Thank you to Rob Kelsey for gathering excellent prayers for us. Using them individually and together will be our duty and our joy.

+ Nicholas Holtam
Bishop of Salisbury and the Church of England's
lead bishop for the environment
Advent 2020

Acknowledgements

From 2002 to 2012, I was the Environment Officer for the Church of England Diocese of Newcastle, and leader of the Diocesan Environment Group. In my early years in the role, it was necessary to argue at Deanery Synods and the like about *why* Christians should care for the earth (to a sometimes sceptical audience). After about five years of the group's existence, we had moved on to address the question of *how* Christians might care for the earth, the previous argument having largely been won. (We were riding a wave of environmental concern, of course, but we helped to swell that wave in our part of the world.)

Those two questions account for the genesis of this book, which began life as a sixteen-page booklet in 2008. It was based on the premise, noted below, that we naturally pray for what we care about, and care more for what we pray about. Although the originating concept was mine, and I collated and edited the prayers (which is how I come to be publishing this book), the idea was formed and developed in the crucible of the group, and the prayers were written by me and other members of the group at the time, namely Peter Dodd, John Harrison, Janis Irvine, Diane Kirkup and Margaret Patterson. I would like to acknowledge and thank them for their past contributions to the work of the group in general, and to the earlier versions of this book in particular, and for their more recent support and encouragement in the production of this expanded version for publication. We have agreed that the proceeds from the sale of this book will be donated to an environmental good cause.

An enlarged version of the original booklet was produced in 2011, with the financial assistance of the Bishop's Council in the Diocese of Newcastle; *Shrinking the Footprint*, the name at the time for the Church of England's environment programme; and the Dioceses of Ripon & Leeds, Wakefield, and York. I remain grateful for the practical assistance of the late Sue Hart and for the encouraging endorsement of the writer Janet Morley.

I wish to thank Victoria Craig and Nick Ellis, who allowed me the use of their flat for the best part of a week, while I prepared most of the text for publication of this book. I would like to thank Mark Nash-Williams, the current Bishop's Adviser on the Environment in the Diocese of Newcastle, for his enthusiastic support and practical advice.

Finally, I would like to thank Bishop Stephen Platten, who encouraged my decision to seek publication; Bishop Nicholas Holtam, for finding time to write the foreword; and all those who have been kind enough to write endorsements.

This book is dedicated to the memory of my dear Mum, Elizabeth Kelsey, and to my dear wife, Kelsie Cox, with thanks for her abiding patience, support and encouragement.

Why should we pray for the earth?

The redemption of all creation

The Bible is clear that God's plan of salvation is for the whole earth, not just human beings. We might think of ourselves as "the crown of all creation", but the natural world is more than a prop for the human race, or a backdrop to the drama of salvation. We might think of ourselves as "the icing on the cake" of creation, but the cake has value in itself, and where would the "icing" be without it? To the extent that human beings abuse and exploit the earth on which we all depend, we are like vandals in an art gallery, despoiling God's beautiful work of creation, or like someone sawing through the branch on which he's sitting.

In the first story of creation in Genesis 1, "God saw that it was good" (Genesis 1:4,10,12,18,21,25) before human beings arrived on the scene. It is worth noting that animals and humankind share the same day of creation (Genesis 1:24–31). Since "God created humankind in his image" (Genesis 1:27), and called on them to "have dominion . . . over every living thing" (Genesis 1:28), we have a God-like, and a God-given, duty to care for the earth and its creatures, as God cares for us. (And, as a vegan for thirty-four years, I can't resist observing that, according to Genesis 1:29–30, a plant-based diet was God's original intention for all living things!)

In the second story of creation in Genesis 2, "the Lord God formed man from the dust of the ground" (Genesis 2:7). The Hebrew word for "man" is *adam*, and the Hebrew word for ground is *adamah* (with a feminine ending), which neatly encapsulates the fact that human beings are part of, and dependent on, mother earth. "The Lord God took the man and put him in the garden of Eden to till it and keep it" (Genesis 2:15), which reminds us again of our calling to care for the earth. And, at Genesis 2:17, God urged restraint on the man, and a limit to his use of the garden's resources.

The Fall from Grace in Genesis 3 had implications, not just for Adam and Eve, but for the earth itself: "cursed is the ground because of you; . . . thorns and thistles it shall bring forth for you" (Genesis 3:17–18). In the modern era in the Western world, human beings have tended to view nature as something wild to be tamed, or an adversary to be subdued, but it was not always so. The insight that human sin has environmental side-effects, and that the wellbeing of humanity is inextricably linked with the wellbeing of the earth, is a recurring theme throughout the Bible (see, for example, Genesis 6:7; Jeremiah 12:4; Hosea 4:1–3; Romans 8:19–21).

The story of Noah and the ark (Genesis 6–9) makes clear that, so far as their salvation is concerned, human beings and animals are "in the same boat" (Genesis 6:20;

8:1). God made his first covenant not just with Noah and his descendants, but also with "every living creature that is with you, the birds, the domestic animals, and every animal of the earth" (Genesis 9:10; see also Genesis 9:12–13,15–17). The common destiny of human beings and animals is reiterated in Isaiah's vision of the Peaceful Kingdom (Isaiah 11:6–9), and at the very end of the story of Jonah (Jonah 4:11).

Job 38–39 and Psalm 104 suggest that God's work of creation is an ongoing process, not a one-off event, and that God is intimately involved with the life, death, and wellbeing of all his creatures. Jesus himself described the way in which God feeds the birds of the air and clothes the lilies of the field (Matthew 6:26,28–30; Luke 12:24,27–28). He also said, "Are not two sparrows sold for a penny? Yet not one of them will fall to the ground unperceived by your Father" (Matthew 10:29), or "not one of them is forgotten in God's sight" (Luke 12:6). Jesus' parable of the mustard seed (Matthew 13:31–2; Mark 4:30–32; Luke 13:18–19) ends with a description of the Kingdom of God as being like a large tree, and the apparently redundant phrase, "the birds of the air come and make nests in its branches" (Matthew 13:32); this might have been a scriptural allusion (e.g. Ezekiel 17:23; 31:6; Daniel 4:12), or a rhetorical ornamentation, but it also suggests that the animal kingdom has a place in God's Kingdom.

The incarnation of Christ, or the embodiment of the Second Person of the Trinity (John 1:14), suggests that physical bodies are important, and signifies that "matter matters to God". The whole of God's creation is sacramental (Romans 1:20), and Christ himself is the quintessential sacrament—an outward and physical sign of an inward and spiritual grace. The bodily resurrection of Christ, likewise, is crucial to a proper understanding of the nature of salvation. Jesus was raised to new life (and ascended to heaven) *in the body*—not the same old body, nor a brand-new body, but his own body, renewed and glorified.

As human beings, we do not—as is often suggested—occupy or inhabit our bodies; on the contrary, our physical bodies are part-and-parcel of who we are. None of us is a person *within* a body; each of us is a person *with* a body. The idea of our spiritual selves "escaping the prison" of our physical bodies is wrong; for the whole person to be redeemed, we must be raised to eternal life—like Christ—in and with our bodies (Romans 8:23). Nothing is discarded; everything is redeemed (or, you might say, recycled!).

The spirituality of heaven is, therefore, not exclusive of the physicality of earth. According to St Paul, "There are both heavenly bodies and earthly bodies. . . . So it is with the resurrection of the dead. [The earthly body] is sown a physical body, it is raised a spiritual body. If there is a physical body, there is also a spiritual body" (1 Corinthians 15:40–44). Jesus himself said that he would drink wine in the Kingdom of Heaven (Matthew 26:29; Mark 14:25; Luke 22:18).

The resurrection of Christ in the body prefigures the re-creation of the universe. "The creation itself will be set free from its bondage to decay and will obtain the freedom of the glory of the children of God" (Romans 8:21). God will "gather up all things in [Christ], things in heaven and things on earth" (Ephesians 1:10). Just as we cannot be saved apart from our bodies, so we cannot be saved apart from the earth. Salvation does not consist in human beings being whisked away to heaven, like passengers from a sinking ship. We will not be saved *from* the earth; we will be saved *with* the earth. Indeed, far from us "going up to heaven" from earth, heaven itself will come down to earth (Revelation 21:2).

Thus, the "new heaven and [the] new earth" of Revelation 21:1 is a *renewed* heaven-and-earth. One thing that Christians might offer the environmental movement is a hopeful vision of the whole earth redeemed by God. This should be, not a source of complacency, but an antidote to despair. Since "all shall be well", the renewal of the earth is assured. Those who love God will not trash the earth in the meantime, in the lazy expectation that God will clear up the mess we've made, but will seek to join with God in his ongoing work of creation and restoration.

The purpose of praying for the earth

Here is not the place for a disquisition on the nature of prayer, or of what we are doing when we pray. Suffice it to say that prayer can be described as "being in the remembered presence of God". (We are always in the presence of God, and, when we remember the fact, then we are praying.) Similarly, intercessory prayer might be described as "remembering others in the presence of God". The idea that prayer persuades God to do something that he wouldn't do otherwise is problematic (though I have experienced occasions of apparently answered prayer). Perhaps the main purpose of prayer is to align our wills with the will of God, just as a piece of metal, in the presence of a magnet, will itself become magnetic.

And so we pray, "Your kingdom come, your will be done, on earth as in heaven." We might pray for the earth in response to any number of environmental crises. But we should pray for the earth on a regular basis, as a matter of course, for the following reasons:

- **Because we love God, and we naturally want to acknowledge God's love for all creation.** Human beings in general, and Christians in particular, tend to be anthropocentric, or human-centred (i.e. self-centred, which helps to explain our environmental crises). When we pray for the earth, then we share in God's delight in, and concern for, the whole of creation.

- **Because we naturally pray for what we care about, and care more for what we pray about**, in a virtuous circle of praying and caring, reflection and action. As we align our wills with God's will, then we are more inclined to do something about it. We form within ourselves—or, rather, God forms in us—a disposition to action. (In the Church of England *Book of Common Prayer*, as each of the Ten Commandments is read out at the beginning of Holy Communion, the congregation responds, "Lord, have mercy upon us, and incline our hearts to keep this law".)

- **Because humanity can be thought of as the "priesthood" of creation, giving voice to the ceaseless praise that the whole creation offers to its Creator.** One thing that distinguishes human beings from other creatures is our use of language, and our ability to articulate the needs of ourselves and others. The words at our disposal are a charism, or gift from God, which enable us to perform God's purpose, for us and all the world. (See Genesis 2:19, where God brought the animals and birds to the first human being, "to see what he would call them; and whatever the man called every living creature, that was its name".) The Church brings the needs of the world before God in intercessory prayer, and we have a duty to pray for the whole earth, not just our fellow human beings.

- **Because human beings need to learn a less self-centred language, in order to re-discover our place within and dependence on the whole of creation.** In the appendix to George Orwell's dystopian vision of the future, *Nineteen Eighty-Four*, "Newspeak" is presented as a language with a continually diminishing vocabulary, designed to limit the freedom of expression, and thus the freedom of thought and capacity for action, of the inhabitants of Oceania. This book aims to do the opposite. Any movement for change involves raising the awareness of those involved. Prayers for the environment are often missing from the intercessions in church on a Sunday morning, perhaps because we struggle to find the right words. But as we practise praying for the earth, we exercise and expand not only our vocabulary, but also our understanding of the issues involved, and our ability to act effectively. We become less anthropocentric, and more inclined to do something about it.

Thus, environmental concerns should be an integral part of the public and private prayers of all Christian people. The use of this book should help us to include such concerns in our corporate and personal prayers.

How to use this book

The prayers in this book are laid out as follows:

- An over-arching Collect for the Earth is provided on page 1.
- A weekly cycle of prayers can be found starting on page 3. The themes of the prayers approximate to the stages of creation in Genesis 1, beginning with Sunday (the first day of creation, and the day of resurrection) and ending with Saturday (the original day of sabbath rest, according to the Hebrew tradition).*
- Two monthly cycles of prayer are provided, starting on page 21 and page 41. The prayers for each month comprise four "weeks" of seven prayers, plus three additional prayers. Within each "week", the themes of the prayers reflect the stages of creation in the weekly cycle.

---

* The theme for Sunday (light and power) combines the first and fourth days of creation (Genesis 1:1–5,14–19), and sometimes includes the idea of enlightenment or learning. The third day of creation (Genesis 1:9–13) is divided between Tuesday (water) and Wednesday (land and vegetation). Although land creatures and humankind share the sixth day of creation (Genesis 1:24–31), the theme for Thursday covers all living creatures, allowing Friday's theme to focus on humanity's relationship to the earth..

- Prayers are provided for the times and seasons of the Church year, starting on page 61, and for various other times, starting on page 69.
- If a prayer on a particular theme is required, it may be found among the list of alternative prayers at the bottom of each page in the weekly cycle, or by referring to the index of prayers starting on page 76.

One benefit of using a prayer cycle is its combination of discipline and variety. Just as a lectionary resolves the question, "What part of the Bible should I read (or preach from) today?", so a prayer cycle resolves the question, "What aspect of the environment should I pray for today?" It helps us to avoid the inclination to do nothing, because we can't decide what to do; or the temptation to keep returning to our favourite topics, while neglecting others; or the tendency to get stuck in a rut, with too narrow a focus.

This book is intended for the use of:

- those who lead the intercessions of the Church.
- groups of Christians whenever they pray together.
- individual Christians in their personal prayers.

The prayers may be used:

- As a supplement to other forms of intercession (see below). The prayers can be adopted unchanged, or

adapted to suit, or as inspiration for praying in
one's own words.

- As prayers in their own right. When used in this
  way, the prayers could include an address to God
  at the beginning, and an ascription to our Lord
  Jesus Christ at the end (as in the Collect for the
  Earth on page 1).

Integrating prayers for the earth within the intercessions of the Church

Most forms of intercession are divided into paragraphs
on particular themes. A typical example may be found
in *Common Worship: Services and Prayers for the Church
of England*, in which the themes follow this pattern:

- The Church in all the world.
- The nations of the world.
- The local community.
- People in need.
- Those who have died.

Prayers for the earth may be used within this pattern, as
shown in Option 1 and Option 2 below. Other options
may be devised for other forms of intercession, as
appropriate.

Option 1

The following paragraph could be inserted either before or after the paragraph on the Church in all the world.

> *Bidding*
> "We pray for the good earth of
>   which we are part."
> *Specific prayer*
> Taken from one of the prayer cycles in this book.
> *Summing-up prayer*
> "Bless the earth and its creatures.
> Give us the wisdom and the will
> to use the resources of the earth to your glory,
> and for the good of all creation."
> *Petition and response*
> As for each of the other paragraphs, e.g.
> "Lord, in your mercy, **hear our prayer**."

Option 2

The paragraph from Option 1 could be combined with the paragraph on the nations of the world.

> *Bidding*
> "We pray for the earth and for the
>   nations of the world."
> *Specific prayer*
> Taken from one of the prayer cycles in this book.
> *Summing-up prayer*
> "Bless and guide *Elizabeth our Queen,*
> give wisdom to all in authority,
> and direct this and every nation
> in the ways of justice, peace and
>   the care of creation,
> that we may honour one another,
> and seek the common good of
>   all that you have made."
> *Petition and response*
> As for each of the other paragraphs, e.g.
> "Lord, in your mercy, **hear our prayer**."

The specific prayers could be chosen:

- according to the day of the month, from one of the monthly cycles (e.g. from Month A in the "odd numbered" months of January, March, May, etc., and from Month B in the "even numbered" months of February, April, June, etc). For a weekly service, this would ensure a variety of subject matter from week to week throughout the year.
- with a particular subject in mind, depending on current concerns or recent events.
- according to personal preference, adapted as necessary or as desired by the intercession leader.

Sensitivity in the choice and/or adaption of a particular prayer might be required, depending on circumstances, such as times of natural disaster.

A Collect for the Earth

All creation worships you, O God,
and we your people bless you for
   the beauty of the earth.
We thank you for making a good
   world in the beginning,
and for making it better by the Word made flesh,
whose death and resurrection in the body reveal
   the promised re-creation of the universe.
May your kingdom come on earth
   as your will is done in heaven
through Jesus Christ our Lord.

**Prayers for the earth
throughout the week**

Sunday

God said, "Let there be light"; and there was light. And God saw that the light was good; and God separated the light from the darkness. God called the light Day, and the darkness he called Night.

Genesis 1:3–5

And God said, "Let there be lights in the dome of the sky to separate the day from the night; and let them be for signs and for seasons and for days and years, and let them be lights in the dome of the sky to give light upon the earth." And it was so.

Genesis 1:14–15

Day 1—Light and power

We thank you for brother sun,
    sister moon and the stars.
We give thanks for the rhythm of
    the days, months and years.
Help us to value both light and darkness.
Grant us wisdom in the use of energy supplies,
and inspiration in the development
    of renewable resources.

Alternative prayers on a related theme from the monthly cycles

Month A

- A1—Times and seasons
- A8—Nuclear power
- A15—Light pollution
- A22—Environmental awareness

Month B

- B1—Renewable energy
- B8—Solar power
- B15—Biofuels
- B22—Education and campaigning

Monday

God said, "Let there be a dome in the midst of the waters, and let it separate the waters from the waters." So God made the dome and separated the waters that were under the dome from the waters that were above the dome. And it was so. God called the dome Sky.

Genesis 1:6–8

Day 2—Air and climate

We thank you for the air that we breathe
    and for the ever-changing skies.
We give thanks for the rhythm of the seasons,
for the warmth of the summer sun and
    the sharpness of the winter frost.
Help us to feel the freshness of the
    breeze upon our faces
and to discern the rainbow of
    hope that you give us.

Alternative prayers on a related theme from the monthly cycles

Month A

- A2—Climate change
- A9—Weather patterns
- A16—Air pollution
- A23—International agreements on climate change

Month B

- B2—Decarbonization
- B9—Clouds and precipitation
- B16—Space junk
- B23—Wind energy

Tuesday

God said, "Let the waters under the sky be gathered together into one place" . . . And it was so. . . . The waters that were gathered together he called Seas. And God saw that it was good.

Genesis 1:9–10

Day 3—Water

We thank you for the life-giving
    waters of the earth,
for the rain that brings refreshment to dry
    land and succour to living things.
Help us to see your peace in the still waters,
your power in the flood and the crashing wave,
your joy in the babbling brook, and your
    timeless presence in the cascading waterfall.

Alternative prayers on a related theme from the monthly cycles

Month A

- A3—The oceans
- A10—Lakes and ponds
- A17—Water pollution
- A24—Flooding

Month B

- B3—Drinking water
- B10—Rivers and streams
- B17—Fish and fishing
- B24—Water shortages

Wednesday

God said, "Let . . . the dry land appear." And it was so. God called the dry land Earth . . . Then God said, "Let the earth put forth vegetation: plants yielding seed, and fruit trees of every kind on earth that bear fruit with the seed in it." And it was so. . . . And God saw that it was good.

Genesis 1:9–12

Day 4—Land and vegetation

We give thanks for the land that sustains
    us, in all its variety and complexity.
We praise you, and pray for high
    mountains and deep valleys,
for fertile plains and desert places, for
    tropical forests and meadow grasslands,
for peat bogs and all fragile habitats.
Make us good caretakers of the land
    on which we all depend.

Alternative prayers on a related theme from the monthly cycles

Month A

- A4—Landscapes
- A11—Trees and woods
- A18—The rainforests
- A25—Gardens

Month B

- B4—The soil
- B11—Plants
- B18—Trees and CO_2 absorption
- B25—Roadside verges

Thursday

God said, "Let the waters bring forth swarms
of living creatures, and let birds fly above the
earth across the dome of the sky." . . . And
God said, "Let the earth bring forth living
creatures of every kind: cattle and creeping
things and wild animals of the earth of every
kind." And it was so. . . . And God saw that
it was good.

Genesis 1:20, 24–25

Day 5—Living creatures

We thank you for the integrity and
    diversity of all living creatures.
Enlarge within us a sense of fellowship
with our brothers and sisters, the animals,
with whom we share the earth, and
    who love the life you give us.
Grant us compassion in our dealings
    with all creatures great and small.

Alternative prayers on a related
theme from the monthly cycles

Month A

- A5—Sea creatures
- A12—Birds
- A19—Farm animals
- A26—Biodiversity

Month B

- B5—Wild animals
- B12—Insects
- B19—Companion animals
- B26—Urban wildlife

Friday

God said, "Let us make humankind in our image, according to our likeness; and let them have dominion over the fish of the sea, and over the birds of the air, and over the cattle, and over all the wild animals of the earth, and over every creeping thing that creeps upon the earth."

Genesis 1:26

Day 6—Human beings and the environment

As you made humankind according
　　to your likeness,
so help us, like you, to see the
　　goodness of creation.
May we see ourselves as part of a greater whole,
remember our duty of care towards the earth,
and live in balance with the natural world.

Alternative prayers on a related theme from the monthly cycles

Month A

- A6—Farming
- A13—Plastics
- A20—Globalization and fair trade
- A27—Organic farming
- A29—Natural disasters
- A30—Transport
- A31—Tourism

Month B

- B6—Tribal peoples
- B13—The built environment
- B20—The waste of war
- B27—Genetic modification
- B29—Pandemics
- B30—Food production and distribution
- B31—Antarctica

Saturday

Thus the heavens and the earth were finished, and all their multitude. And on the seventh day God finished the work that he had done, and he rested on the seventh day from all the work that he had done. So God blessed the seventh day and hallowed it, because on it God rested from all the work that he had done in creation.

Genesis 2:1–3

Day 7—Rest and restraint

We thank you for sabbath moments
    in the busyness of our lives.
May we have time to stand and stare, to
    reflect on the beauty of creation,
and to appreciate the many blessings of this life.
Help us to be satisfied with enough,
and to live within our means in
    relation to the earth.

Alternative prayers on a related theme from the monthly cycles

Month A

- A7—Reduction of consumption
- A14—Sustainable development
- A21—Waste reduction
- A28—Access to the countryside

Month B

- B7—Re-use and recycling
- B14—Food waste
- B21—Litter
- B28—Mental wellbeing

**Prayers for the earth
throughout the month**

Month A

Week 1

A1—Times and seasons

We thank you for the rhythm
    of times and seasons.
Make us more aware of our
    changing environment,
in our parks and gardens, in the
    countryside and by the sea.
Help us to appreciate all that is special
    about the present time of year,
and to live our lives in keeping
    with the natural order.

A2—Climate change

Grant us wisdom and foresight in
    the face of global warming.
We pray for those affected by rising sea
    levels and extreme weather conditions.
Help us, in reducing our own carbon footprint,
to play our part in reducing our
    country's carbon emissions.
Grant us the personal and political
    will to make a difference.

A3—The oceans

In the beginning your Spirit swept
    over the face of the waters,
and the mighty oceans of the world
    reflect your majesty and glory.
May the wind and the waves remind
    us of your creative power.
In your mercy protect all who work
    and travel on the sea,
and preserve the teeming life of
    coral reef and ocean depth.

A4—Landscapes

We thank you for the physical
    landscape that surrounds us,
for hills and dales, woods and
    fields, roads and paths.
We thank you, too, for the inner landscape
    of our thoughts and feelings.
As the environment forms us,
may we form an environment that is
    beneficial to all your creatures.

A5—Sea creatures

We give thanks for the diversity of
    life in the seas of the world,
for the grandeur of whales, the beauty of
    fish and the intricacy of coral reefs.
We pray for an end to commercial whaling.
May we prevent the pollution
    of our seas and rivers,
and maintain the integrity of our
    blue and beautiful planet.

A6—Farming

We pray for all who make their
    living from the land.
Help them to contend with the
    vagaries of the weather
and the variability of the economic climate.
Defend them from disease, and help them
    to protect the animals in their care.
We ask for your blessing on all farm
    shops and farmers' markets.

A7—Reduction of consumption

We thank you for the resources of the world,
and for the many goods available to us.
Grant us wisdom and restraint in our
    spending and consumption,
help us to be satisfied with enough,
and make us mindful of the value of simple gifts.

Week 2

A8—Nuclear power

We pray for wisdom in our response to the
    apparent promise of nuclear power.
Help us to weigh up the benefits
    of a carbon-free supply
and the dangers of radioactive waste.
Bless those who work in the nuclear industry,
and those involved in the transport
    of nuclear material.

A9—Weather patterns

We give thanks for the weather
    in its infinite variety.
We thank you for the rain and the snow,
the winter winds that blow and the
    warmth of the summer sun.
As we read the signs of the ever-
    changing weather,
help us also to understand the signs
    of the times in which we live.

A10—Lakes and ponds

We thank you for the precious
  waters of the earth,
for the beauty of a lake or reservoir, and
  the micro-ecology of a pond,
for their mirror stillness and murky depths.
Help us to preserve their life-giving qualities,
and the flora and fauna that
  depend on their presence.

A11—Trees and woods

As we give thanks for the company of trees,
may we stand in awe of how lovely they are.
Remind us of their power to heal the minds
  and spirits of a fallen humanity.
Teach us to protect the trees,
and preserve ancient woodlands
  for generations to come.

A12—Birds

As we consider the birds of the air,
we praise you for their beauty, and
    the sweetness of their singing.
We give thanks for the part they play
    in the spreading of seeds.
As we admire their ability to fly above the earth,
give us the wings of faith to see your
    kingdom in the world around us.

A13—Plastics

We pray for a reduction in the use of
    plastic throughout the world,
and for a solution to the prevalence of plastic
in rivers and oceans, and in marine
    animals and birds.
Help us to reduce our reliance on plastic,
in bottles and bags, and
    consumables of all kinds.

A14—Sustainable development

We give thanks for all the blessings of this life,
and remember those less
    fortunate than ourselves.
Help us, who have enough, to live more
    simply, that others may simply live.
As we borrow the earth from our children,
may we not take for ourselves
    more than we can return.

Week 3

A15—Light pollution

We thank you for the lights of the natural world
    and for the lights of human ingenuity.
Help us to appreciate the variety of
    light and the value of darkness.
Grant us wisdom in the use of light,
so as not to pollute the awesome
    beauty of the night sky.
May we continue to see the stars, and
    wonder at our place in the universe.

A16—Air pollution

We give thanks for the air in which we
    live and move and have our being.
As the wind blows where it wills,
grant all nations wisdom in dealing
    with cross-border pollution.
Help us to value the atmosphere
    as a life-giving resource,
and to make amends when we pollute its
    purity, for the sake of all your creatures.

A17—Water pollution

We give thanks for the sustaining
  power of the waters of the earth.
Forgive us when we are wasteful and
  take clean water for granted.
Forgive us when we are careless and
  contaminate the waters of life.
Help us to nurture the living water of our faith,
and to protect the springs and
  rivers of the world.

A18—The rainforests

In the rainforests of the earth you reveal
  the diversity of your creation.
Help us, in wonder, to care for your forests,
and in so doing to protect and
  strengthen the lungs of the earth.
Bless native peoples and forest
  dwellers, and help us with them
to learn the lesson of our shared
  belonging and interdependence.

A19—Farm animals

We give thanks for your promise of salvation,
   to us and to every living creature.
We praise you for the wild animals of the earth
and ask your blessing on all beasts of
   burden and food production.
Forgive us when we are careless
   of your creatures,
and convict us of our cruelty.

A20—Globalization and fair trade

We pray for guidance in the global economy.
We give thanks for the diversity of foods
   and products available to us,
and remember those who provide
   the goods that we enjoy.
Help us to support the principles of fair trade,
and to practise it in our daily lives.

A21—Waste reduction

We give thanks for the material things of life.
May we use them wisely, and dispose
   of them thoughtfully.
In business and as individuals, help
   us to reduce what we waste,
and to remember that, as the
   earth is our only home,
nothing is altogether thrown away.

Week 4

A22—Environmental awareness

May we taste and see how good the world is.
Grant us sensitivity to the touch of rain and sun.
Help us to hear the sounds of animals and birds,
and smell the scents of fresh air and flowers.
Raise our awareness of the beauty
    and susceptibility of the earth.

A23—International agreements on climate change

We give thanks for the ways in which
    countries have come together
to address the global crisis of climate heating.
We praise you for the efforts of individuals
who have raised our awareness, and
    changed hearts and minds.
May we, together, create the political
    climate for effective action.

A24—Flooding

We pray for your blessing on those
  affected by flooding.
Help us, as a society, to plan housing
  developments carefully,
provide effective flood protection,
and manage the environment to reduce
  flooding in the first place.
May we work with, not against,
  the natural way of things.

A25—Gardens

We give thanks for all that grows in our gardens,
for grass and flowers, fruit and
  vegetables, insects and animals.
May our gardens be a delight to the
  eye, an oasis for nature,
and a balm to the soul.
Draw us closer, in our gardens, to
  your heart for all creation.

A26—Biodiversity

Through the process of evolution,
you have created a multiplicity of life on earth.
Make us mindful of the rich variety
    of life in all its fullness.
Bless all endangered species of flora and fauna,
and help us to protect the biodiversity
    of the natural world.

A27—Organic farming

We thank you for the gifts of the land,
and pray your blessing on the farmers
    who make those gifts available to us.
We give thanks for those who respect the earth
and who seek to maintain its natural balance.
We pray for an encouraging market for
    those who produce organic food.

A28—Access to the countryside

We pray for those who live in and
   visit the countryside.
We give thanks for rights of way
   and the right to roam.
May we protect the green spaces
   round our towns and cities,
and make the countryside available to everyone,
for the refreshment of urban dwellers, and
   the wellbeing of rural communities.

Additional prayers

A29—Natural disasters

Help us to understand the powers
    at work in your creation,
in the storm and flood, in the
    earthquake and landslide.
Grant us wisdom in choosing where
    and how to live and work.
Help us to support our local
    and global neighbours
when natural disasters devastate their lives.

A30—Transport

We give thanks for the ability to travel widely,
in this country and around the world.
Help us to discern the true cost of transport,
to use bikes and buses where possible,
    and share lifts with other people.
May we tread lightly on the
    earth, wherever we go.

A31—Tourism

You have granted us the ability to explore
    your world as never before.
In all our journeys, open our eyes
    to the beauty of the earth,
give us peace of mind and relief from stress.
May we also travel thoughtfully,
and reduce the imprint of humanity
    upon a fragile ecology.

Month B

Week 1

B1—Renewable energy

We give thanks for the power
    of the sun above us,
for the warmth of the ground beneath us,
for the freedom of the wind around us,
for the strength of waves and
    the rhythm of tides.
Grant us ingenuity and inspiration in the
    development of renewable energy.

B2—Decarbonization

We pray for the transformation of
    our carbon-based economies.
Give wisdom and inspiration to
    our leaders and investors,
to plan a way forward, and to
    force the pace of change.
Help us to play our part in steering the world
towards a more sustainable future.

B3—Drinking water

We offer thanks for easy access to safe drinking
    water, sufficient to meet our needs.
We remember those who lack
    clean water supplies,
and those who walk for miles to
    fetch water for their families.
Bless the engineers and development agencies
who seek to make a difference in their lives.

B4—The soil

We give thanks for the soil beneath our feet,
which supports and sustains all
    the creatures of the earth.
As the soil provides nourishment for us,
may we nourish the soil on
    which our lives depend.
Make us mindful that good soil brings
    forth a multitude of benefits.

B5—Wild animals

We give thanks for the wild animals of the earth,
with whom we share the gift of life.
Grant us a sense of kindred with
    our fellow living creatures.
Make us the friend of helpless things,
and help us to prevent all cruelty and neglect.

B6—Tribal peoples

We pray for tribal peoples across the world,
for their human rights to be respected,
and for their land rights to be recognized.
We pray for an end to logging and
    mining on tribal lands,
and for the safeguarding of the
    livelihoods of all tribal societies.

B7—Re-use and recycling

Grant us inspiration in waste reclamation,
and in the re-use and recycling of resources.
Bless those who work in the recycling industry,
and those who work in charity shops.
Help us to find new homes for unwanted items.

Week 2

B8—Solar power

You make your sun rise on the
    evil and on the good.
It goes forth from the end of the heavens,
    and runs to the very end again,
and there is nothing hidden from its heat.
As the sun gives life and light to the earth,
may we harvest its power for the
    good of all your children.

B9—Clouds and precipitation

We bless you for the clouds that
    float across the sky,
and for sleet and snow, and soft refreshing rain.
As they come down from above,
    and water the earth,
make us responsive to your word
    and fruitful in good works.
Pour upon us and all creation the
    continual dew of your blessing.

B10—Rivers and streams

We praise you for the rivers and
    streams of the world,
for their moving, yet abiding, presence,
    which is ever old and ever new.
We give thanks for all good and beautiful things
that flow from the past into the future.
Help us to maintain our rivers and
    streams of living water.

B11—Plants

You make grass to grow upon the hills,
    and green plants to serve our needs.
We give thanks for valleys
    standing thick with corn,
and for the flowers of the field in
    meadows and margins.
May a rich variety of species be clothed in glory,
and provide refuge and food for
    all manner of animals.

B12—Insects

The insects of the field are yours,
and we praise you for the part they play
in the fertilization of crops and flowers.
Help us to preserve the populations
    of bees and other insects.
May we kill not the moth nor butterfly,
    for your tender mercy's sake.

B13—The built environment

We pray for the environment
    of our own making,
for town and countryside, buildings and fields.
We give thanks for the ingenuity and
    artistry of engineers and architects.
Bless those who maintain the fabric of our cities,
and those who protect our green
    and pleasant land.

B14—Food waste

We thank you for the food we eat,
and we remember before you those
    who sometimes miss a meal,
as they struggle to make ends meet.
Help us, in our homes, and as a nation,
    to avoid wasting good food.
May we buy as much as we need,
    and share whatever we can.

Week 3

B15—Biofuels

Teach us wisdom in our use of
    the earth's resources.
Grant us inspiration in our search
    for alternatives to fossil fuels.
Give guidance to those involved in
    the development of biofuels.
Help us to find the right balance
between the need for food and the need for fuel.

B16—Space junk

We give thanks for the ethereal
    beauty of outer space,
and praise you for the purity of
    the spangled heavens.
As we extend the sphere of human activity,
keep us from cluttering the sky and
    spreading space debris.
We pray for the restrained and sustainable
    use of satellite technology.

B17—Fish and fishing

We praise you for fish and all
    that live in the waters,
and pray for the protection of
    endangered species.
Bless those who make their living from the sea,
and help us to strike a wise balance
    between fishing for food
and preserving the integrity of
    the marine environment.

B18—Trees and CO_2 absorption

We praise you for the natural
    capacity of trees and plants
to absorb and store carbon dioxide.
As we seek to respond to the climate emergency,
we give thanks for those who plant and
    preserve our woods and trees.
May the leaves of the trees help
    the healing of the nations.

B19—Companion animals

We ask for your blessing on our pets
    and animal companions.
We praise you for their abiding presence,
for their loyalty and lack of judgement,
for their simplicity and humility.
May we care for them, as you care for us.

B20—The waste of war

We remember with sorrow the waste of war:
the waste of human and animal life,
the waste of resources, and the
    devastation of the environment.
Help us to find better ways of resolving conflict,
and to work together for the peace and
    wellbeing of all living things.

B21—Litter

We pray for a world free of litter,
for clean streets and pavements,
    country roads and lay-bys.
Change the hearts and minds of those
    who drop litter without a care,
and those who deliberately flout the law.
Help us to protect the natural world from
    the selfishness of the human race.

Week 4

B22—Education and campaigning

Bless teachers and scientists, as they
    reveal the wonders of creation.
May all children have the chance to explore
    and experience the natural world.
Grant wisdom and effectiveness to
    environmental campaigners
as they seek to change hearts and minds,
and raise awareness of our need to
    respect and protect the earth.

B23—Wind energy

The Holy Spirit, like the wind, sweeps
    over the face of the earth,
giving life to the world and all its creatures.
We pray for those who seek to harness
    the power of the wind,
and for those concerned about the
    siting of wind turbines.
Grant wisdom and integrity to those who
    form opinions and make decisions.

B24—Water shortages

As the world gets warmer, we pray for
    those affected by the scarcity of water.
Guide our response to the droughts,
    and help us resolve the disputes
caused by lack of rainfall and depleting supplies.
May the nations of the world work together
to preserve and share the resources you give us.

B25—Roadside verges

We pray your blessing on grass verges
    and roadside nature reserves.
We praise you for the animals that make
    their home in quiet corners,
for the wild flowers that flourish,
and for the birds, bees and butterflies
    that find refuge and refreshment.
May we treasure all small areas of
    outstanding natural beauty.

B26—Urban wildlife

We pray for the animals and plants
　　that live in cities and towns:
the birds and bats that share our buildings,
the weeds that appear at the edge of pavements,
and the creatures that find a haven
　　in our parks and gardens.
Help us, so far as we are able, to live and let live.

B27—Genetic modification

From the beginning you have created
　　plants and creatures of every kind.
Help us, who have dominion
　　over every living thing,
to respect the integrity of every species.
Enlighten the debate about genetic modification.
May we not be blinded either by
　　science or by prejudice.

B28—Mental wellbeing

We give thanks for the power
    of the natural world
to slow our hectic pace of life and
    soothe our ragged nerves.
Make us ever mindful of your
    presence in creation,
bless those who live with mental illness,
and help us all to find our rest in
    the beauty of the earth.

Additional prayers

B29—Pandemics

In the plagues and pandemics that
    afflict the human race,
may we realize our vulnerability,
and learn the lesson that we are part-
    and-parcel of a fragile ecology.
Take from us a desire to dominate,
    lest it be our undoing,
and teach us a new humility in
    our relation to the earth.

B30—Food production and distribution

We give thanks for the food we eat,
and for the people who bring it to our plate.
Help us to find the right balance between
    buying and selling locally,
and supporting our global neighbours
    in developing countries.
Grant us a greater appreciation of
    good food in due season.

B31—Antarctica

We praise you for Antarctica,
earth's only continent without a
   native human population.
We give thanks for the Antarctic Treaty
and pray for the continuing prohibition
   of mineral mining and nuclear waste disposal.
Bless the penguins and seals, and the
   research work of visiting scientists.

The times and seasons
of the Church year

C1—Advent

Grant us the expectant hope that
    this damaged earth
will be healed of its pain and
    restored to wholeness.
Come among us to judge the
    world with righteousness
and to change both hearts and minds.
Turn us away from exploitation
    towards the nurture of creation.

C2—Christmas

We praise you for the Word made flesh,
for heaven come down to earth and
    earth raised up to heaven.
Being found in human form, Christ embodied
    his love for the material world.
Help us, like him, to care for the
    earth of which we are part
and to give substance to our concern
    for the environment.

C3—Epiphany

In the manifestation of Christ, you use the
    things of the earth to reveal your glory.
As we take gold from the ground,
make frankincense and myrrh
    from the gums of trees,
and turn grapes and water into wine,
grant us wisdom to use the resources of
    the earth for the good of all creation.

C4—The Presentation of Christ in
the Temple (Candlemas)

Purify our hearts and pierce them
    with the knowledge
of the ill we have done to your precious creation.
Lighten our understanding, to see your
    salvation in the wellbeing of all things.
Bless the work of scientists, as they
    bring enlightenment to the world,
and direct their efforts towards
    the healing of the earth.

C5—Lent

In this time of abstinence and reflection,
help us to confront our weaknesses
    in the wilderness
and perceive our place in your great universe.
We give thanks for the green shoots of your
    presence in the waste places of the world.
May we learn new habits of holy living
    in our relationship with the earth.

C6—Mothering Sunday

We remember before you our mothers,
and all those who nurture and
    care for other people.
We give thanks for Mother Church, and
    those who have raised us in the faith.
We pray for mother earth, who has given us life.
May we, and all her children, respect her
    in return, and care for her wellbeing.

C7—Holy Week (or Holy Cross Day)

We give thanks for the gift of life
   from the dust of the ground,
and for the new life that rises
   from the dust of death.
As all things hold together in Christ,
who died on the cross and rose to new life,
may we and all creation be released from
   decay and renewed in his image.

C8—Easter

At this time of new life, we praise you
   for the power of your creation.
The bodily resurrection of Christ proclaims
   the promised re-creation of the universe.
In the renewal of the natural world,
in the blossom of trees and the birth of animals,
help us to celebrate your many gifts of
   springing hope and fresh beginnings.

C9—The Ascension

The bodily ascension of Christ reveals the
    significance of the material world.
As he was exalted to heaven, may we
    in heart and mind also ascend,
so as to view the environment from
    a kingdom perspective.
Help us to see the world as you see it
and care for the earth because you love it.

C10—Pentecost

We ask for the wisdom of the Holy Spirit
to know our place in the universe and
    to play our part in creation.
May the Spirit of truth help us to speak the
    truth about the earth and its creatures.
Inspire and revitalize those who
    care for the environment,
and nurture in all people a deeper
    engagement with the living world.

C11—The Holy Trinity

We praise you for the diversity and
    harmony of all that you have made.
The myriad forms of life on earth are knit
    together and depend upon each other.
The particles and planets of the universe
    have a singular coherence.
As we ponder the mystery of
    the Trinity in Unity,
may we play our part in the integrity of creation.

C12—All Saints (or any saint)

Your saints in every age proclaim your
    glory in the highest heavens
and see your beauty in the smallest
    of your creatures.
May we, inspired by the depth of their devotion,
discern your sacred presence
    in the material world,
and offer you our lives in praise and service.

C13—All Souls

You sustain the fabric of the universe
    and preserve all souls in life.
As you encompass heaven and earth,
    so we, in life and death,
are united with those who have died and
    with everything you have made.
Help us so to reverence all living things
that we come to a deeper knowledge
    of your care for all creation.

C14—Christ the King

We remember, with shame,
that we have sometimes abused the
    high dominion of humankind,
so that the song of the earth has
    been a groan of travail.
May we learn from the humility
    of Christ our King,
and worship him in harmony with
    the whole of creation.

Prayers for the earth at other times

D1—Christian Unity

*The Week of Prayer for Christian Unity is kept from
18 to 25 January in the Northern Hemisphere.*

As we rejoice in the diversity of
    the universal Church,
may we also rejoice in the rich
    diversity of your creation.
Make us mindful of our unity in Christ.
Help us realize your plan for the fullness of
    time, to gather up all things in him,
things on heaven, and things on earth.

D2—Plough Sunday

*Plough Sunday is traditionally celebrated
on the First Sunday of Epiphany.*

As we mark another turning in
    the agricultural year,
may we turn over a new leaf in our
    dealings with the earth.
Bless the plough and the soil, the
    farmer and the consumer.
Make us mindful of our interdependence
and ready to give a fair return
    for all that we receive.

D3—Saint Cuthbert (or any northern saint)

The Festival of St Cuthbert is usually celebrated on 20 March.

We give thanks for St Cuthbert and
   all the northern saints.
As they lived in harmony with
   their environment,
may we also acknowledge your
   presence in the natural world.
Open our eyes to the wonders of creation,
and help us to delight in the rhythm
   of times and seasons.

D4—Rogationtide

*Rogationtide is traditionally celebrated on the Sixth
Sunday of Easter and/or the following three weekdays.*

We ask for your blessing on the land
   and on our local environment.
We pray for those who work with the earth
to bring forth its fruits in due season.
Grant us wisdom in conserving
   the earth's resources,
and sustaining the land that gives us sustenance.

D5—Christian Aid

Christian Aid Week is usually held during the second week of May.

As changing weather patterns
  and rising sea levels
affect the poorest peoples of the world,
we confess our contribution to the
  threat of climate change.
Bless the work of Christian Aid,
and guide us and all the world into the
  ways of justice and of peace.

D6—Sea Sunday

Sea Sunday is usually held on the second Sunday of July.

We remember those who go
  down to the sea in ships
and ply their trade in great waters.
May they see your works and
  wonders in the deep,
and praise you for your goodness.
Strengthen and sustain those in peril,
  and bring them safely home.

D7—Lammas

Lammas Day is traditionally celebrated on 1 August, in thanksgiving for the first fruits of the wheat harvest.

We acknowledge with gratitude the
   mysterious gift of ripened grain
and the simple goodness of a loaf of bread.
We give thanks for the seed, the
   soil, the rain and the sun,
and for the joyful hope of a bountiful harvest.
Give to all people the nourishment they
   need for their souls and bodies.

D8—The Blessed Virgin Mary

The Festival of the Blessed Virgin Mary is usually celebrated on 15 August or 8 September.

In her obedience to your will,
the Blessed Virgin Mary gave
   substance to your plan
for the salvation of the universe.
May we, like Mary, realize your sacred
   presence in our ordinary lives,
and may your will be done on
   earth as it is in heaven.

D9—Saint Francis

The Festival of St Francis is celebrated on 4 October.

We give thanks for St Francis,
and for all those who, following his example,
care for the earth and all its creatures.
Help us to respect the natural world
    and to protect all living things,
for they are precious in your sight.

D10—Harvest

*Harvest Thanksgiving is usually celebrated
in late September or early October.*

We give thanks for the fruit of the
    earth that sustains human life.
May we be good stewards of the
    resources you give us
and resist the temptation to take
    everything for ourselves.
Give us the grace to live wisely and generously,
by nurturing the world we occupy and
    by sharing the bounty we enjoy.

D11—Angels

The Festival of St Michael and All Angels is on 29 September.

We praise you for all that you have created,
in heaven and on earth, visible and invisible.
As we contemplate the mystery of
   the company of angels,
may we recognize our place in
   the order of the universe.
May your holy angels help us to
   sustain and renew the earth.

Index

The following is not an exhaustive index of subjects, but lists the titles of the prayers in alphabetical order.

Lightning Source UK Ltd.
Milton Keynes UK
UKHW020837111121
393781UK00005B/214

9 781789 591354